A New True Book

SPACE COLONIES

By Dennis B. Fradin

CHILDRENS PRESS ®

CHICAGO

PHOTO CREDITS

NASA—2, 6, 13, 14, 19, 23, 25, 27, 29, 33, 34, 37, 38, 39, 41, 44

Historical Pictures Service, Chicago—4 (2 photos)

© M. Cole—7 (right)

© Jim Rowan—7 (left)

© Milt and Joan Mann—8 (2 photos), 10 (2 photos), 43

Space Studies Institute—Cover, 16, 21, 31

© Tony Freeman—35 (2 photos)

Cover—Spacecraft leaving for moon from space colony

Artist's concept of a possible space colony, a cylinder 19 miles long and 4 miles in diameter. Housing several hundred thousand people, the colony would be solar powered and would rotate on its axis once every 114 seconds to create an earthlike gravity. The bridge shown would be the size of the San Francisco Bay Bridge.

Library of Congress Cataloging in Publication Data

Fradin, Dennis B.
    Space colonies.

    (A New true book)
    Includes index.
    Summary: Discusses building colonies in space—why, where, and how—and the benefits to Earth.
    1. Space colonies—Juvenile literature. [1. Space colonies] I. Title.
TL795.7.F73 1985   629.44'2   85-7722
ISBN 0-516-01273-8   AACR2

TABLE OF CONTENTS

Plymouth Colony in 1622 (above).
The first winter in Plymouth Colony (below) was a harsh one.

WHAT ARE
SPACE COLONIES?

A colony is a settlement built by people who have gone to live in a new place. The Plymouth Colony was built by English settlers called Pilgrims who arrived in America in 1620.

Many other colonies have been built *on* our

Glass flasks hold communities of shrimp, algae, and microorganisms that can make their own food, oxygen, and water. The flasks have been sealed for over a year. Such experiments help scientists understand what kinds of life support systems humans would need in a space colony.

planet Earth. In the next hundred years people will probably build colonies far *above* Earth, in outer space. These will be called space colonies.

WHY BUILD COLONIES IN SPACE?

Earth is a lovely blue and green planet. It has air and water. It has land for growing crops and grazing animals. Why would anyone want to leave our planet and live in a space colony?

Overcrowding in American suburbs (above)
and in Tokyo, Japan (right),
the world's most populated city

One reason is that Earth
is getting overcrowded. In
1950, Earth's population
was 2.5 billion. In 1980, it
reached 4.4 billion. By the
year 2000 the population
may be 7 billion. Such a
large population will mean

less food and less room for each person. One solution would be for some of the people to live in space colonies.

Another reason is that Earth has only so much oil and natural gas. We are already running low on these fuels. One day we may run out of them altogether.

In space, there are almost endless sources of energy. The heavenly

Decades of open-pit iron mining (above) and strip mining for coal (right) are gradually using up the earth's mineral resources.

bodies in space contain many minerals. Energy from sunlight, called solar energy, can produce great amounts of electricity. One day soon scientists may be able to mine those

minerals and gather
sunlight in space for use
on Earth.

Where will the people
who mine the minerals
live? Where will those who
gather the solar energy
live? The answer,
according to many
scientists, is in space
colonies!

WHERE SHOULD THE COLONIES BE BUILT?

There are two ways to create a space colony. The first is to build it on another heavenly body—on the moon or Mars, for example. The other way is to build a place in space where people can live.

There are problems with building colonies on other heavenly bodies. Because

Colonies on the moon may look like this.

of the pull of gravity, it
would take large amounts
of fuel to land or take off
from the surface of planets
or large moons. Also, other
heavenly bodies are too
hot or too cold for people
to live on without heavy

Skylab

suits for protection. And no other nearby planets have air for people to breathe.

Most scientists think it would be better to build the colonies in man-made worlds. In a small way, this already has been done. *Skylab,* launched in 1973,

was a man-made
spacecraft in which
astronauts lived for months.
A space colony would
have to be much larger
than *Skylab,* but not that
much different.

The first space colonies
would probably be placed
somewhere between Earth
and the moon. As the
distant parts of the solar
system are explored,
colonies might be built
farther and farther from

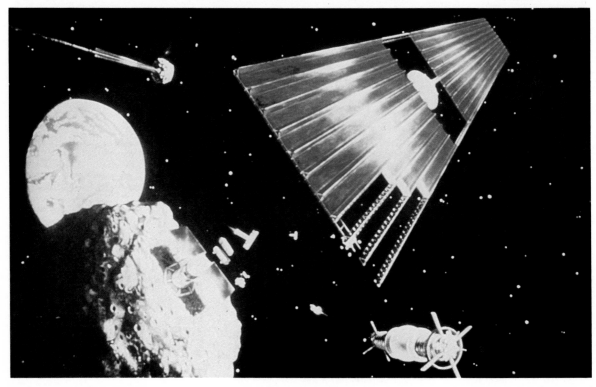

Mining an asteriod (left) and building a solar power satellite (blue structure)

Earth. For example, if people mine minerals from asteroids someday, space colonies may be built in the asteroid belt between Mars and Jupiter.

BUILDING A SPACE COLONY

Many scientists predict that the first space colonies will be ball-shaped objects. Each colony would weigh about 200,000 tons. Because of its size, a colony would not be launched in one piece from Earth. Instead, it would be put together in space, piece by piece.

Many raw materials will be needed to build the first colony. Among them are metals, glass, rock, soil, concrete, water, and air. It would be very expensive to send all these things from Earth. So scientists would get many of them from beyond Earth.

The moon would be a big source of raw materials. It has large amounts of aluminum and titanium, good metals for

Scientists grew plants in samples of moon soil brought back by astronauts.

building. It has much silicon, which is used to make glass and concrete. The moon also has vast stretches of ten-foot-deep soil. Tests of lunar soil show that plants can grow in it quite well.

Water is formed by combining oxygen and hydrogen. Oxygen supplies are trapped in moon rocks. Hydrogen sent from Earth could be combined with oxygen from the moon to produce the space colony's water.

How would all these materials from the moon reach the space colony? Gravity, the force that holds objects down to a heavenly body, is not as strong on the moon as it

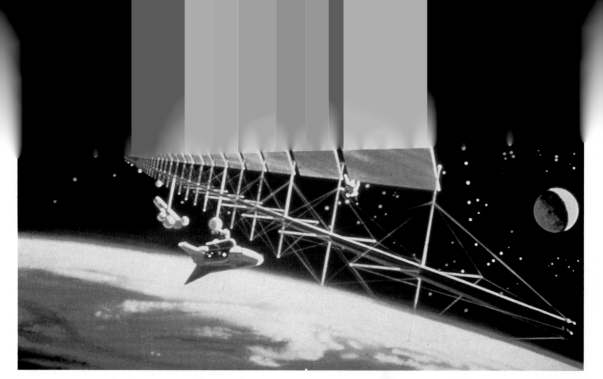

A mass driver being built in space, with
a space shuttle bringing construction materials

is on Earth. So it would
not be very difficult to
launch raw materials from
the moon. Scientists have
built models of a launcher
to do the job. It is called
a mass driver.

Of course, some materials and machinery would have to come from Earth. Earth would also send hydrogen to the colony.

Building the first space colonies would be exciting. Spacecraft from Earth would be constantly arriving with supplies. Meanwhile, moon miners would be busy launching lunar materials toward points in space. From

Residents might view this scene from inside their homes in a wheel-shaped space colony.

there cargo ships would carry these materials to the colony.

Once construction of the first colony began, it could be finished in a few years. Then it would be time for the tenants to move in— the first space colonists!

LIFE IN A SPACE COLONY

Up to ten thousand people could live in a colony one third of a mile wide. In many ways, their life would be very much like ours on Earth.

A space colony would not have much gravity of its own. Without gravity the colonists would float around like astronauts in space. Scientists have found a way to solve this problem. They would

Ten thousand people could live on this mile-wide rotating space settlement. It features a river with shores of lunar sand.

create artificial gravity in the colony by spinning it.

If you've ever been on a merry-go-round, you know the feeling of being pushed to the outside as you spin. This is caused by centrifugal force. The space colony will spin like a merry-go-round. Inside the colony everything will be forced toward the walls. Then the walls will really be like the floor for the colonists.

The colony will have buildings on the inside. They will be made of glass, metal, and concrete taken from moon materials. Some buildings will be

This huge space colony, 250,000 miles from Earth, would be built almost entirely from minerals mined on the moon.

apartments where people will live. Others will be schools, houses of worship, and businesses.

Outside the buildings there will be sidewalks and parks. There may even be man-made streams and hills. There may be streets, but there won't be any cars. The first colonies won't be big enough for cars to be needed.

Sunlight will be reflected through the colony's windows by mirrors. Because most people like having day and night, the windows will have shutters that are closed for about eight hours of each twenty-four-hour period. This will be the "nighttime"

Space farms could grow food for a space colony.

when most of the space
colonists will sleep.

The colony will grow its
own food. To avoid using
inside space, there will be
farm areas attached to the
outside of the colony. Such
crops as wheat, corn,

soybeans, and rice will be more plentiful than crops grown on Earth. They will receive sunlight twenty-four hours a day. Pigs, cows, chickens, and other livestock will also be raised to provide food for the colonists.

The colonists will work at many jobs. Some will grow crops and raise livestock. Others will be doctors, teachers, and storekeepers. Many will work at producing power for Earth.

The colonists will also have time for fun. They will be able to read books on home computers and watch TV programs from Earth. Sports such as

Astronauts' living quarters, built from the liquid-hydrogen container of a space shuttle external fuel tank

soccer, tennis, football, and baseball also will be popular.

Baseball pitchers will have one big advantage up there. Because the colony is spinning, a ball pitched in a straight line won't go straight. Pitches will have a natural curve of ten inches or more!

Space colonists also could enjoy special sports. At the center of the colony the artificial gravity will be

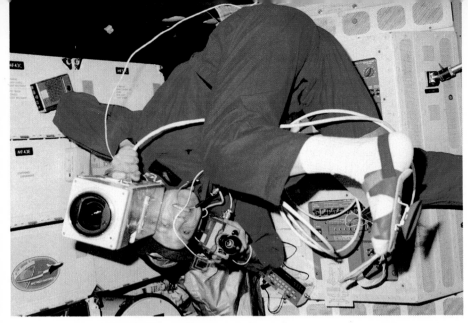

Astronaut Thomas K. Mattingly floats in zero-gravity conditions on the space shuttle *Columbia*. The suction cups on his shoes can be used to hold him in place.

very weak. Colonists could play zero-gravity basketball in special padded rooms. Flying will be possible with small wings strapped to the arms. In low gravity a diver could jump off a diving board and have ten seconds to twist and turn before reaching the water.

Solar panel on a space shuttle collects solar energy.

BENEFITS FOR EARTH

In the future there probably will be a network of sunlight collectors scattered in space. These will "catch" large amounts

These solar panels near Nesperih, California, provide electricity for a hundred people.

of sunlight and beam it to
Earth as radio signals. On
Earth, these signals will be
converted to electricity.

Some scientists think
that sunlight collectors
could one day provide
Earth with all the electricity

it needs. This would solve our energy problems and also clean up our environment.

People in the space colonies will build and maintain the sunlight collectors. Spacecraft will take them to the collectors where they will do their work. Miners who work on the moon and on other heavenly bodies will also travel from the colonies by

Optical flying-spot scanners measure the position and speed of sacks of lunar material on their way to the mass driver before being launched into space.

spacecraft. Some of the rare materials they mine may also be sent to Earth.

Outside each colony there may be small work areas with no gravity. Zero gravity conditions are good

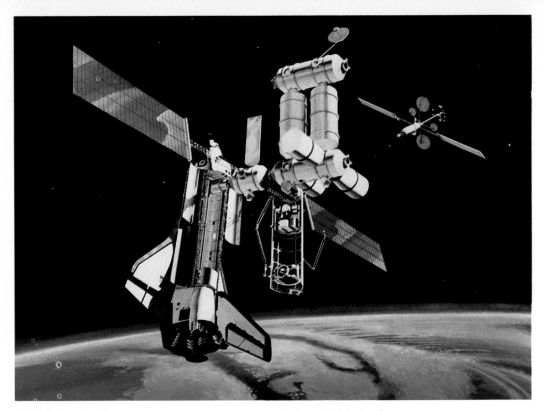

A space shuttle on its way to a space colony might dock
at a space station to drop off supplies.

for making products, such
as crystals for electronics
equipment and some metals
and medicines. These
products will also be
exported for use on Earth.

WHEN WILL IT HAPPEN?

No one knows exactly
when the first space
colony will be built. Many
scientists think there will
be colonies in space by
the year 2010 or so. As

This cut-away view of a space colony shows
an interior that rotates to create gravity and
an outer shell that keeps out cosmic rays.

time passes, each colony will depend less and less on Earth.

If the first space colonies are successful, larger colonies will probably be built. Some of these colonies will be shaped like long tubes. Eventually there may be colonies that hold 100,000 or even a million people. Such large colonies would help solve Earth's population problem.

Giant mirrors beam sunlight into space colonies during "daytime."

Many scientists believe that sometime in the next century there may be more people living in space colonies than on Earth. When Earth can depend on factories and power plants in space, much of our planet may become fields and forests again.

Pollution could someday wipe out life on our planet.

HOPE FOR THE FUTURE

Our Earth, which is four-and-a-half billion years old, is not going to last forever. About five billion years from now the sun will expand. It will burn up all life on our planet.

43

Five billion years is a long time from now. But one day the end will come. Will all human beings be destroyed in the flames of the sun?

Those people of the future will have a way to save themselves. They will be able to leave Earth and go to live on other worlds—in space colonies!

WORDS YOU SHOULD KNOW

artificial (ar • tih • FISH • il) — made by people rather than occurring naturally

asteroids (AST • er • oidz) — rocky objects located between Mars and Jupiter

astronauts (AST • roh • nawts) — space travelers

billion (BILL • yun) — a thousand million (1,000,000,000)

centrifugal force (sen • TRIH • fi • gil FORSS) — the force that pushes things outward on a spinning body

colony (KAHL • uh • nee) — a settlement built by people who have gone to live in a new place

electricity (ih • lek • TRIH • sih • tee) — a kind of energy that provides us with heat, light, and power

gravity (GRAV • ih • tee) — the force that attracts all things to other things and holds objects down to heavenly bodies

lunar (LOO • ner) — pertaining to the moon

mass driver (MASS DRY • ver) — a special kind of electric motor that can launch materials off the moon

million (MILL • yun) — a thousand thousand (1,000,000)

minerals (MIN • er • elz) — useful substances taken from the ground

oxygen (OX • ih • jen) — a substance that is found in both air and water

population (pop • yoo • LAY • shun) — the number of people who live in a place

Skylab (SKY • lab) — a space station launched by the United States in 1973

solar (SO • ler) — pertaining to the sun

solar system (SO • ler SISS • tim) — the sun and all objects that orbit it

space(SPAISS) — the region beginning about one hundred miles from earth

space colony(SPAISS KAHL • uh • nee) — a colony built somewhere in outer space

ton(TUN) — two thousand pounds

INDEX

About the Author

Dennis Fradin attended Northwestern University on a partial creative writing scholarship and graduated in 1967. He has published stories and articles in such places as Ingenue, The Saturday Evening Post, Scholastic, Chicago, Oui, *and* National Humane Review. *His previous books include the Young People's Stories of Our States series for Childrens Press, and* Bad Luck Tony *for Prentice-Hall. In the True book series Dennis has written about astronomy, farming, comets, archaeology, movies, the space lab, explorers, and pioneers. He is married and the father of three children.*